PREFACE

For intelligent identification and solution of problems arising during operation and maintenance of aircraft, it is essential that aviation personnel have a clear understanding of the physical principles involved. A study of aerodynamic principles help aviators remain aware of the capabilities and limitations of their aircraft. Maintenance personnel will be more efficient if they know about aircraft materials and structural principles. Most of all, the safety staff should have a good basic knowledge of these subjects.

It is the purpose of this publication to review the basic physics principles and the minimum mathematics required to solve simple equations in the further study of aerodynamics, aircraft structures, and related subjects. This text is used in the USAF Flying Safety Officer Course at the University of Southern California.

The information is presented in the form of a programmed learning text. It is designed for self study. Answers to the simple questions are found in the right hand margin. These should be covered with a two inch strip of opaque paper. The students should answer the question, then slide the opaque paper strip down to check their answer. Additional problems will be found at the end of each chapter.

It is recommended that the student use a pocket calculator to solve the problems. If possible, the calculator should be capable of extracting square roots and performing trigonometric functions.

Requests for permission or further information should be addressed to C. E. Dole, 1503 Franklin Avenue, Redlands, CA 92373, telephone (714) 793-9768.

                                                Charles E. Dole, EdD
                                                The Safety Center
                                                Institute of Safety and Systems Management
                                                University of Southern California
                                                February, 1974

MATHEMATICS AND PHYSICS
FOR AVIATION PERSONNEL

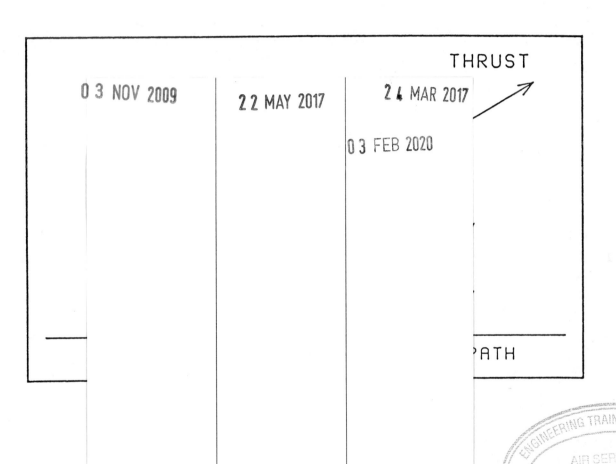

CONTENTS

CHAPTER ONE	ALGEBRA REVIEW	1
CHAPTER TWO	TRIGONOMETRY REVIEW	13
CHAPTER THREE	FORCES and VECTORS	23
CHAPTER FOUR	MOMENTS	31
CHAPTER FIVE	NEWTON'S LAWS OF MOTION	43
CHAPTER SIX	LINEAR MOTION	55
CHAPTER SEVEN	CURVILINEAR MOTION	67
CHAPTER EIGHT	ENERGY	73
CHAPTER NINE	WORK and POWER	79
CHAPTER TEN	FRICTION	87

USE BOOKMARK TO COVER ANSWERS

IN THE RIGHT HAND MARGIN

CHAPTER ONE

ALGEBRA REVIEW

INTRODUCTION

The title of this chapter is a misnomer. We are going to review only a small portion of algebra. We will concentrate on the manipulation of equations. Usually an equation is given in one form only, for instance, $F = ma$.

If we know the values of F and m and want to find the value of a. We would have to rewrite the original equation. This is the type of manipulation that we will cover in this chapter.

The mathematically minded student may now skip to the practice problems at the end of the chapter. If these problems can be solved, nothing will be gained by covering the intervening material.

1. Algebra is a form of mathematics in which letters and symbols are used to represent numbers and to state rules. In some equations, e.g. $x - x = 0$, a letter can be replaced by any number and the equation will still be true. In others, the letter can only be replaced by one number. In the equation $5 + x = 8$, the value of x can only have a value of _____.   3

2. In other equations, the value of one letter depends upon the value of another letter. In the expression $a + b = 10$, the value of a _____ upon the value of b.   depends

3. In the study of aviation there are many problems that require the use of physics formulas which we will be investigating in this book. We will concentrate on solving algebraic expressions called _____.   equations

4. Equations are mathematical sentences in algebra. They say two things are equal. Here is a simple equation:

$$7 + y = 12$$

To solve this equation we must find some number that makes this sentence true. If you replace the letter y with any number except ____, the equation will be false.

    5

5. Most algebraic equations are more complex than the above example. Algebra has methods of changing difficult equations into simple ones. We will concentrate on _____ equations.

    solving

6. To help in solving equations, we rearrange or transpose the equation. Transposing means _____ the letters and numbers in an equation.

    rearranging

7. The basic rule in transposing is:

WHATEVER YOU DO TO ONE SIDE OF THE EQUATION,

YOU MUST ALSO DO TO THE OTHER SIDE.

If you subtract a number or a letter from the left side of an equation, you must _____ the same quantity from the right side of the equation.

    subtract

8. Think of an equation as a balance scale. Whatever you do to one side of the scale, you must also do to the other side, or it won't be _____.

    balanced

9. In order to find the values of the letters in (solve) an algebraic equation, you must have enough information given so that you can find the unknown quantities. We say that you are solving for those quantities. For example in the equation $a + b = 10$, we could solve for either a or for b. If you wanted to solve for a, you would transfer all other letters and numbers from the left side of the equation to the right side except ____ . a

10. To solve for a you would remove b by subtracting it from *____   both
side(s) of the equation.   * the left, the right, both

11. Remember that you must treat both side alike. After you subtract "b" from both sides, the equation is: $a + b - b = 10 - b$.

The $+b$ and the $-b$ on the left side cancel, leaving the equation in the transposed form as: $a =$ _____ . $10 - b$

12. To find the value of a in the above equation, you must be given (or be able to find) the value of b. If the value of b is 2, you can substitute 2 for b and find $a =$ ____ .   8

13. The basic processes of algebra are the same as those of arithmetic. These are: addition, subtraction, multiplication, and division. The results of addition are called the SUM. The results of subtraction ate called the DIFFERENCE. Subtraction is the opposite of addition. To transpose a letter or number that is added to others in an equation we must ____   subtract
the letter or number. Letters and numbers are called SYMBOLS.

14. But we must be sure that we also subtract the symbol from the

_____ side of the equation.   other

15. USING SUBTRACTION TO SOLVE AN EQUATION.

Rule: To maintain the equality, equal symbols must be subtracted from **both** sides of an equation.

Solve the following equations for u.

(a) $u + 8 = 11$ $u = $ _____.   3

(b) $u + x = 6$ $u = $ _____. $6 - x$

(c) $u + ab = 6a$ $u = $ _____. $6a - ab$

(d) $mx = u + rv$ $u = $ _____. $mx - rv$

(e) $u + v = 90°$ $u = $ _____. $90° - v$

16. USING ADDITION TO SOLVE AN EQUATION.

Rule: To maintain an equality, equal symbols must be added to both sides of an equation.

Solve the following equations for v.

(a) $v - 19 = 21$ $v = $ _____.   40

(b) $v - x = 8$ $v = $ _____. $8 + x$

(c) $v - 2x = 4x$ $v = $ _____. $6x$

(d) $4 = v - 16$ $v = $ _____.   20

(e) $v - a^2 = b^2$ $v = $ _____. $b^2 + a^2$

Note: The product of a number "n" with itself is denoted n^2. $5^2 = (5)(5) = 25$

17. We can multiply one side of an equation by any symbol, provided we

_____ the other side by the same symbol.   multiply

So $g = h + 6$ is the same as $2g = 2h + 12$.

Multiplication and division are opposites.

18. In a fraction such as $\frac{a}{b}$, a is called the *numerator* and b is called the

denominator. If we want to remove a symbol that is in the denominator of a fraction from one side of an equation we must multiply both sides of the equation by that symbol. If we want to solve $\frac{W}{Y} = z$ for W, we want to

eliminate the denominator of the fraction. So we multiply _____ side(s)   both

by _____. The Ys on the left side cancel and $W =$ _____.   Y, zY

19. Multiplication symbols can be shown in several different ways:

A x B, A · B, (A)(B), and AB, all mean the same thing. They mean that

A and B are _____ together.   multiplied

20. **USING MULTIPLICATION TO SOLVE AN EQUATION.**

Rule: To maintain an equality, both sides of an equation must be multiplied by the same symbol(s).

Solve the following equations for y.

(a) $\frac{y}{3} = 5$ $y =$ _____.   15

(b) $\frac{y}{x} = 40$ $y =$ _____.   40x

(c) $\dfrac{6}{y} = 1$ $y = \underline{}$.   6

(d) $1 = \dfrac{4}{y}$ $y = \underline{}$,   4

21. If two or more symbols are multiplied together we can eliminate the unwanted symbols by dividing _____ side(s) of an equation by the unwanted symbols.   both

22. If we want to solve the equation $AB = CD$ for A, we want to remove the undesired symbol, ____ . B Since A and B are multiplied together, we must use the opposite process which is _____.   B   division

23. So we divide _____ side(s) of the equation by _____.   both "B"

24. The result is $A = \underline{}$. $\dfrac{CD}{B}$

25. **USING DIVISION TO SOLVE AN EQUATION.**

Rule: To maintain an equality, both sides of an equation must be divided by the same symbol(s). Note: Dividing by zero is not allowed.

   Solve the following equations for x.

(a) $7x = 35$ $x = \underline{}$.   5

(b) $33 = 11x$ $x = \underline{}$.   3

(c) $3xy = 9$ $x = \underline{}$. $\dfrac{3}{y}$

(d) $6xy = 36y^3$ $x = \underline{}$. $6y^2$

26. RECAP

In an equation where the symbols are joined together by + or - signs, we remove the unwanted symbols thus:

To remove a positive symbol _____ the symbol from each side.	subtract
To remove a negative symbol _____ the symbol to each side.	add

In an equation where the symbols are multiplied together or divided by each other, we remove the unwanted symbols thus:

If the symbol is in the denominator of a fraction, we _____ both sides by it.	multiply
If the symbol is in the numerator of a fraction, we _____ both sides of it.	divide

27. There are two "short-cut" methods, which do not seem to obey the rule that each side must be treated in the same manner. If you examine the results, however, you will see that the answers are the same.

Short-cut rule No.1: To remove an unwanted symbol that is added to or subtracted from other symbols, move it directly over to the other side of the equation and **change its sign.**

Example: Solve this equation, $F + 16 = D$, for F.

Solution: $F = D - 16$.

Try the short-cut method on questions 15 and 16.

Try the short-cut method on questions 15 and 16.

28. Short-cut Rule No.2: To remove an unwanted symbol that is in the numerator or denominator of a fraction **that stands alone** on one side of an equation, cross multiply **without changing its sign**. By cross multiplying we mean, moving the symbol from the numerator on one side of the equation and multiplying it to the denominator on the other side. The reverse process is also valid. Remember that **all** of the symbols on the opposite side must be multiplied (or divided) by the cross multiplied symbols. Example: Solve the equation $ABC = 2x$ for A. Solution: $A = \dfrac{2x}{BC}$

Solve this equation for A. $\dfrac{A}{z} = xy$ $A = $ _____ . xyz

Try the short-cut method on questions 20 and 25.

29. Here is an example of using all four of the above processes.

Solve $\dfrac{2y}{3} - 4 = \dfrac{y}{4} + 6$ for y.

STEP 1: (multiplication) It is easier to deal with whole numbers rather than fractions. IF we multiply both sides by the smallest number into which both denominators can divide evenly (lowest common denominator), we can eliminate the fractions. This number is 12.

The new equation is: $8y - 48 = 3y + 72$.

STEP 2: (addition) Add 48 to each side. The equation is: $8y = 3y + 120$.

STEP 3: (subtraction) Subtract $3y$ from each side. The equation is: $5y = 120$.

30. When multiple operations are required to solve an equation, first remove undesired fractions (STEP 1). Next isolate the term(s) containing the unknown symbol by transposing added or subtracted symbols (STEPS 2,3). Finally transpose undesired multiplied or divided symbols (STEP 4).

Example: $\frac{5z}{2} - 7 = 3$.

1. multiply by 2: $5z - 14 = 6$

2. transpose -14: $5z = 6 + 14$

3. add: $5z = 20$

4. cross multiply 5: $z = 4$.

Solve the following equations for z.

(a) $2z + 7 = 11$ $z = $ _____ .   2

(b) $\frac{z}{2} + 6 = 4$ $z = $ _____ .   -4

(c) $2z - \frac{x}{2} = \frac{6x}{5}$ $z = $ _____ . $\frac{17x}{20}$

(d) $\frac{5x}{z} + 3 = 18$ $z = $ _____ . $\frac{x}{3}$

The symbol $\sqrt{a}$ is used to denote a number that when multiplied by itself yields a. It is called the square root of a. So $(\sqrt{a})(\sqrt{a}) = a$.

CHAPTER ONE - MORE PRACTICE PROBLEMS

1. Given: $V = at + V_o$ (Note: V_o is one symbol) Solve this equation for:

a. $V_o = \underline{}$. $\hfill V - at$

If $V = 50$, $a = 4$, and $t = 5$, then $V_o = \underline{}$. $\hfill 30$

b. $t = \underline{}$. $\hfill \dfrac{V - V_o}{a}$

If $V = 100$, $a = 5$, and $V_o = 50$, then $t = \underline{}$. $\hfill 10$

c. $a = \underline{}$. $\hfill \dfrac{V - V_o}{t}$

If $V = 60$, $t = 6$, and $V_o = 30$, then $a = \underline{}$. $\hfill 5$

2. Given: $s = \tfrac{1}{2}at^2 + V_o t$ Solve this equation for:

a. $V_o = \underline{}$. $\hfill \dfrac{s - \tfrac{1}{2}at^2}{t}$

If $s = 1000$, $t = 10$, and $a = 4$, then $V_o = \underline{}$. $\hfill 80$

b. $a = \underline{}$. $\hfill \dfrac{2(s - V_o t)}{t^2}$

If $s = 500$, $t = 10$, and $V_o = 40$, then $a = \underline{}$. $\hfill 2$

3. Given: $2as = V^2 - V_o^2$. Solve this equation for:

a. $V = \underline{\qquad}$. $\sqrt{V_o^2 + 2as}$

If $V_o = 10$, $a = 2$, and $s = 11$, then $V = \underline{\qquad}$. 12

b. $V_o = \underline{\qquad}$. $\sqrt{V^2 - 2as}$

If $V = 10$, $a = 2$, and $s = 9$, then $V_o = \underline{\qquad}$. 8

4. Given: $KE = \frac{1}{2}mV^2$ and $m = \frac{W}{32}$. (note: KE is one symbol)

Substitute the value of the second equation into the first and solve for:

a. KE in terms of V^2 and W. $KE = \underline{\qquad}$. $\dfrac{WV^2}{64}$

If $W = 3200$ and $V = 100$, then $KE = \underline{\qquad}$. 500,000

b. $V = \underline{\qquad}$. $\sqrt{\dfrac{64KE}{W}}$

If $KE = 3{,}200{,}000$ and $W = 3200$, then $V = \underline{\qquad}$. 252.98

11

5. Given: $L = C_L qS$ and $q = \dfrac{\sigma V^2}{295}$. (note: C_L is one symbol, σ is the Greek letter "sigma".)

Substitute the value of the second equation into the first and solve for:

a. $L = \underline{\qquad}$. $\hfill \dfrac{C_L \sigma V^2 S}{295}$

If $C_L = 0.5$, $\sigma = 0.8$, $V = 200$, and $S = 500$, then $L = \underline{\qquad}$. $\hfill 32{,}000$

b. $V = \underline{\qquad}$. $\hfill \sqrt{\dfrac{295L}{C_L \sigma S}}$

If $S = 250$, $C_L = 0.5$, $\sigma = 0.8$, and $L = 20{,}000$, then $V = \underline{\qquad}$. $\hfill 223.6$

c. $C_L = \underline{\qquad}$. $\hfill \dfrac{295L}{\sigma V^2 S}$

If $S = 250$, $\sigma = 1.0$, $L = 25{,}000$, and $V = 200$, then $C_L = \underline{\qquad}$. $\hfill 0.7375$

CHAPTER TWO

TRIGONOMETRY REVIEW

1. The word TRIGONOMETRY means triangle measurement. But, today, trigonometry means much more than the measurement of triangles. Mathematicians have studied the relations between the three sides of triangles and have expressed these relationships as RATIOS. In our study we will deal with the properties and application of trigonometric _____.

                                                                                                     ratios

2. Triangles can be classified by the size of the angles. All three angles of any triangle add up to 180°. If one of the angles of a triangle is greater than 90°, the triangle is called an OBTUSE triangle. The sum of the two smaller angles in in an obtuse triangle is *_____ 90°.

   *equal to, greater than, less than

                                                                                                    less than

3. A triangle that has no angle which is 90° or greater is an ACUTE triangle. In an acute triangle, the sum of any two angles is *_____ 90°.

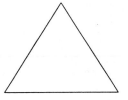

*equal to, greater than, less than

                                                                                                  greater than

4. A triangle which has one angle exactly equal to 90° is called a RIGHT triangle.

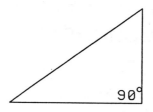

In a right triangle the sum of the two smaller angle is *_____ 90°.

*equal to, greater than, less than

equal to

ONLY RIGHT TRIANGLES WILL BE DISCUSSED IN THIS CHAPTER.

5. In a right triangle we can easily find one of the small angles if we know the other by _____ the known angle from 90°.

subtracting

6. In a right triangle, if we find one of the small angles to be 57°, the other small angle will be _____.

33°

7. The longest side of a right triangle is the side opposite to the 90° angle and is called the HYPOTENUSE.

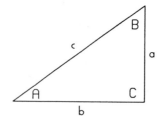

In the right triangle shown here, side _____ is the hypotenuse.

c

14

Capital letters are usually chosen to represent angles and lower case letters are used to represent corresponding opposite sides.

8. If we are talking about angle A in this triangle, the side that is across from angle A is called the opposite side and is side "a". The side that is closest to angle A is called the adjacent side. It is side _____ . b

9. Side "c" in the above triangle is always called the _____ and is never the opposite or adjacent side. hypotenuse

10. The Greek mathematician Pythagoras (in 586 B.C.) proposed the theorem which bears his name. It is known as the Pythagorean Theorem: "FOR A RIGHT TRIANGLE, THE LENGTH OF THE HYPOTENUSE SQUARED IS EQUAL TO THE SUM OF THE SQUARES OF THE LENGTHS OF THE OTHER TWO SIDES." $c^2 = a^2 + b^2$.

If we know _____ side(s) of a right triangle, we can find the third side. two

11. If we know that a = 3 and b = 4. What is the value of c? _____ . 5

12. If we know that c = 10 and a = 6. What is the value of b? _____ . 8

13. So far we have found out that, in a right triangle, if we know the size of one of the small angles, we* _____ find the size of the other small angle.   *can or can not.   can

Also, if we know the length of two of the sides, we *_____ find the length of the other side.   *can or can not   can

14. Find the unknown angle or side in this right triangle.

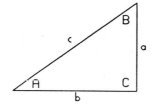

a. A = 23°, B = _____ .   67°

b. B = 40°, A = _____ .   50°

c. a = 4, b = 5, c = _____ . $\sqrt{41}$

d. b = 2, c = 6, a = _____ . $\sqrt{32}$

e. c = 7, a = 4, b = _____ . $\sqrt{33}$

15. At the beginning of this chapter we mentioned that there are certain mathematical ratios between the three sides of a triangle and the angles of the triangle. Consider the right triangle shown here.

There are six possible combinations of ratios formed by any two sides and the angle _____ .

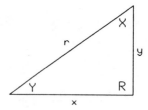

Y

16

16. The Sine of angle Y (Sin Y)

is the ratio of the side opposite

to angle Y to the hypotenuse.

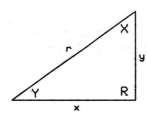

$Sin\ Y = \dfrac{opposite\ side}{hypotenuse}$ or $Sin\ Y = $ ——.

$\dfrac{y}{r}$

17. **The Cosine of angle Y** (abbreviated as Cos Y) is the ratio of the side adjacent to angle Y to the hypotenuse.

$Cos\ Y = \dfrac{opposite\ side}{hypotenuse}$ or $Cos\ Y = $ ——.

$\dfrac{x}{r}$

18. The Tangent of angle Y (abbreviated as Tan Y) is the ratio of the side opposite to angle Y to the adjacent side.

$Tan\ Y = \dfrac{opposite\ side}{adjacent\ side}$ or $Tan\ Y = $ ——.

$\dfrac{y}{x}$

19. The other three trigonometric ratios are merely reciprocals of the first three and are not required to completely "solve" a triangle.

Cotangent (COT) is the reciprocal of Tangent.
Secant (SEC) is the reciprocal of Cosine.
Cosecant (CSC) is the reciprocal of _____.

Sine

20. These ratios are called TRIGONOMETRIC FUNCTIONS. Let's see how we can use them to solve right triangles.

If we want to find the value of "x" in this

right triangle. We must select a trigonometric

function which includes x and the known

side. _____ .

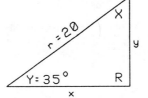

r

21. $$Sin\ Y = \frac{y}{r},\quad Cos\ Y = \frac{x}{r},\quad Tan\ Y = \frac{y}{x}$$

The <u>only</u> function containing x and r is _____ .   Cos Y

22. So $Cos\ Y = \frac{x}{y}$ must be solved for x. Multiplying both sides of the

equation by r, we get x = _____ .   r(Cos Y)

23. Substituting the known values for r and Y, we get

x = ____ Cos ____ .   20(Cos 35°)

24. Using your scientific pocket calculator, press 35, then cos,

multiply by 20, and find x = _____ .   16.383041

25. If we want to solve for "y", we look for a function which contains y

and two known quantities.

If we had not previously solved for x, the only function that we could use

would be _____ .   Sin Y

26. Solve $Sin\ Y = \frac{y}{r}$ for y.

y = ____ .   20(sin 35°)

27. Using your scientific pocket calculator, press 35,

then sin, multiply by 20, and find y = _____ .   11.471529

28. Of course, once we had solved for x (in question 24) we could use the

value of x as a known quantity.

This is not recommended as any error in finding x will be carried along.

29. Once we had solved the above triangle for x, we could find y by using the Pythagorean theorem $r^2 = x^2 + y^2$.

$y = $ _____. $\sqrt{r^2 - x^2}$

30. We can also solve a right triangle if _____ sides are known.   two

31. To solve the right triangle shown, we must find the value of side _____.

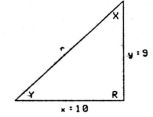

   r

and the angles _____ and _____.   Y and X

32. Can we solve for side r without using the trigonometric functions? _____.   yes

33. The _____ theorem will let us solve for r directly   Pythagorean

34. r = _____ (what letters?) $\sqrt{y^2 + x^2}$

   r = _____ (what numbers?) $\sqrt{9^2 + 10^2}$

   r = _____ (number)   13.453624

35. Once we have found the value of r, then we can solve for angle Y by using *_____ of the functions.   any one
   *only one, two, any one

36. Now that we know all of the sides, we can find angle Y by any one of the functions.

$Sin\ Y = \dfrac{y}{r} = \dfrac{9}{13.453624}$ $Y = $ _____.   41.987213°

$Sin\ Y = \dfrac{x}{r} = \dfrac{9}{13.453624}$ $Y = $ _____.   41.987212°

$Tan\ Y = \dfrac{y}{x} = \dfrac{9}{10} = 0.9$ $Y = $ _____.   41.987212°

37. Again it is recommended that the function involving the two given values be used to prevent carry-over errors.

For the above problem, the _____ function would be less likely to produce such a mistake.

Tan

SUMMARY OF CHAPTER TWO FORMULAS

$a^2 + b^2 = c^2$

$A + B = 90°$

$Sin\ A = \dfrac{a}{c}$ $\quad$ $Sin\ B = \dfrac{b}{c}$

$Cos\ A = \dfrac{b}{c}$ $\quad$ $Cos\ B = \dfrac{a}{c}$

$Tan\ A = \dfrac{a}{b}$ $\quad$ $Tan\ B = \dfrac{b}{a}$

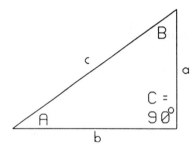

CHAPTER TWO - MORE PRACTICE PROBLEMS

1. In this right triangle fill in the blanks.

   a = ____

   Sin A = ____

   Cos A = ____

   Tan A = ____

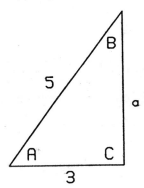

   4

   0.8

   0.6

   1.33333

2. In this right triangle solve for x and y.

   Tan 60° = ____/100

   x = ____ Tan 60°

   x = ____

   Cos 60° = ____/y

   y = 100/____

   y = ____

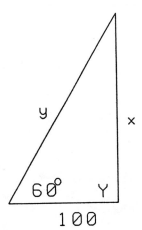

   x

   100

   173.20508

   100

   Cos 60°

   200

3. An airplane takes off and flies at a fixed angle of 9° to the horizontal ground. When it is at an altitude of 400 feet above ground level, Find:

   a. The horizontal ground distance the plane has flown. _____   2525.50 feet

   b. The actual distance through the air that the plane has flown. _____   2556.98 feet

21

4. An airplane travels 15,000 ft. through the air at a uniform angle of climb, thereby gaining 1830 ft. of altitude. Find the angle of climb. _____.   7.01°

5. An airplane is on an instrument final approach and is flying at 150 knots. The GCA glide slope is 3° from the horizontal. What should the rate of climb indicator read (feet per minute)? (Hint: To obtain ft/min multiply knots by 101.3)   R/C = _____   -795.24 fpm

6. In plotting a wreckage diagram you find that it isn't possible to take measurements that form a right triangle, as shown below.

It is still possible to find the distance u and angle A by using right triangle formulas. Find them. u = _____ ft., A = _____.   u = 1307.67

A = 36.59°

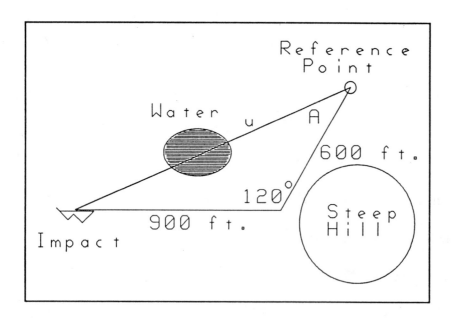

CHAPTER THREE

FORCES and VECTORS

1. A FORCE, F, is defined as a push or pull on an object and is measured in pounds (lb.). In the picture below, the jet THRUST is a _____ force acting in the direction of flight.

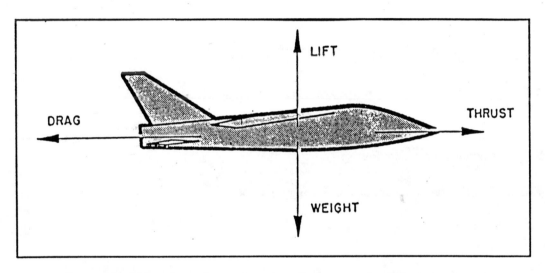

2. Aerodynamic DRAG is a force that acts in the opposite direction to _____ and resists the motion of thrust on the airplane.   thrust

3. The LIFT on the airplane is a force acting perpendicular to the flight path. If the plane is flying horizontally without gaining or losing altitude, it *_____ equal to the weight of the plane.   is

   * is or is not

23

4. The WEIGHT of the plane is a force that acts toward the center of mass of the earth. Regardless of the plane's attitude the weight acts as a force in *_____ direction.   the same

*the same or a different

5. Each of these forces of thrust, drag, lift and weight is measured (in the USA) in the units of _____.   pounds

6. PRESSURE, P, is the effect of a force on a unit area of a body. The formula, $P = \dfrac{Force}{Area}$, shows this relationship. Pressure is measured in pounds per square inch (psi) or pounds per square foot (psf).

If there is a pressure of 5 psi acting on a surface of 6 square inches, the total force on the surface is _____ lb.   30

7. If an airplane weighs 20,000 lb. and has a wing area of 500 square feet, the lifting pressure on the wings will average _____ psf, if the plane is in level flight. This is called the plane's **WING LOADING**.   40

8. A SCALAR quantity is one that has size only. It is not necessary to know anything about directions when considering scalars. The quantities of mass, time and temperature are examples of _____.   scalars

9. In adding (or subtracting) scalar quantities we use simple arithmetic. If a man drove his car 20 miles on one day and 55 miles on the next day, he drove a total of _____ miles.   75

10. A quantity that has both size and direction is called a VECTOR QUANTITY or simply a VECTOR. Force is a *_____.   vector

   * scalar or vector

11. If we said that the man in question 9 drove his car 20 miles EAST, then his movement would be a vector quantity, because both distance and direction are specified.

In physics, we say the car's DISPLACEMENT was _____ miles to the   20

_____ of where it started.   east

12. We could draw a diagram of the path of the car like this:

   ———————A————▶   The arrow, A, represents the vector

quantity. The head of the arrow indicates the direction that the car traveled and the length of the arrow represents the _____ that the car has traveled.   distance

13. If the car moved from A to B by way of C then it *_____ moved in one straight line. The car *_____ undergone two separate displacements.

   *has or has not

                                                                                     has not

                                                                                     has

14. Vectors can be added or subtracted by using graphic methods. The car in question 13 above is traveling from A to C (vector X) and then from C to B (vector Y). This shows an example of how vector Y was _____ to vector X. The resultant is vector _____.   added, Z

15. Forces, velocities, distances and accelerations are examples of vectors. It is important to know in which _____ they act as well as their _____.   direction, size

16. If an airplane is flying west at 200 knots true airspeed and has a 50 knot tailwind, its ground speed is _____ to the _____.   250 kts, west

17. Speed is a scalar quantity but VELOCITY includes the direction of movement. Velocity is a _____ quantity.   vector

18. The vector picture of the airplane in 16 above is shown here.

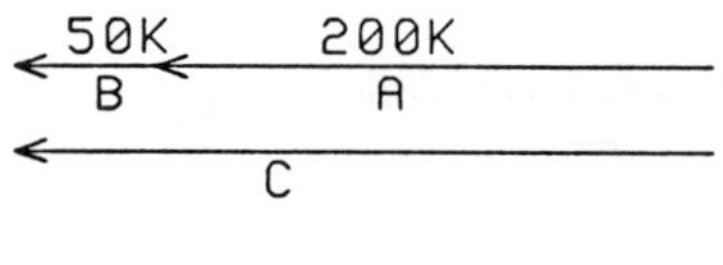

```
   50K      200K
◄───── ◄──────────
   B        A
◄──────────────────
          C
```

Vector A represents the _____ of the airplane through the air.   velocity

Vector B shows the _____ of the wind.   velocity

Vector _____ represents the resultant ground velocity.   C

This is vector addition of vectors A and B.

19. Suppose that an airplane is flying north at 100 knots and encounters a cross wind from the west of 30 knots. Here is the vector picture. The velocity of the plane through the air is represented by vector _____.

The vector which represents the wind velocity is _____.   Y

The ground speed is represented by the *_____ of vector Z.   length

and the track over the ground is represented by the *_____ of vector Z.   direction

*direction or length

20. An example of vector subtraction is seen in the following:

An aircraft is flying to the east at 200 knots and encounters a 40 knot headwind.

Vector P represents *_____ velocity of the airplane.   air

The wind velocity is represented by vector _____.   Q

The resultant vector, R, represents the *_____ velocity of the airplane.   ground

* ground or air

21. In graphically adding two vectors we can use the TRIANGLE LAW:

(1) Place the tail end of either vector at the origin (the point where the vector acts).

27

(2) Place the tail end of the second vector at the arrow end of the first vector. The resultant vector of the two vectors is drawn from the origin to the arrow end of the second vector. This is shown below.

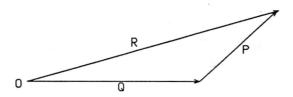

The first vector is Q. The second vector, ____, is added as described above.   P

The resultant vector is ____.   R

22. Find the resultant vector of the following pairs of vectors.

a.
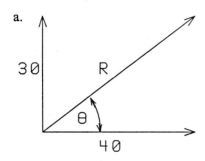

R = ____.   50

θ = ____.   36.9°

b.
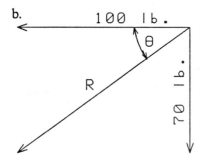

R = ____   122.1 lb.

θ = ____.   35°

28

23. A vector can be resolved (replaced) by two component vectors that are perpendicular to each other. If these two component vectors had acted on a body, instead of the original vector, the result would have been *_____.   the same

*the same or different

24. If we draw a line from this missile vertically to the horizontal line, we have constructed a right triangle. The side of the triangle opposite

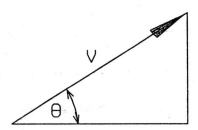

to the angle is the vertical component of the vector V and represents the *_____ velocity of the missile.   vertical

The side of the triangle that is adjacent to angle represents the horizontal component of vector V and represents the *_____ velocity of the missile.   horizontal

*horizontal or vertical

25. We can find the values of the components by solving the right triangle. If we know the velocity of the missile is 2000 feet per second and the angle, θ, is 40°, the equations are:

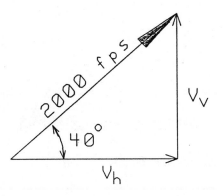

$$\sin 40° = \frac{V_v}{2000} \quad \text{and} \quad \cos 40 = \frac{V_h}{2000}.$$

or $V_v =$ _____ fps and $V_h =$ _____ fps.   1285.6, 1532.1

29

26. Find the horizontal and vertical components of the following vectors.

a.

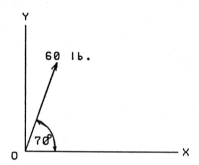

    Fx = _____ lb.   Fy = _____ lb.   20.5, 56.4

b.

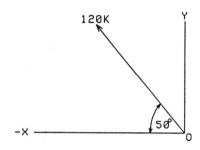

    Vx = _____ .   Vy = _____ .   -77.1, 91.9 K

27. An airplane is towing a sail plane (glider). The tow rope is 20° below the horizontal and has a tension force of 300 pounds exerted on it by the airplane. Find the horizontal drag of the sail plane and the amount of lift that the rope is providing.

    Drag = _____ lb.   Lift = _____ lb.   281.9, 102.6

CHAPTER FOUR

MOMENTS

1. A lever is a rigid bar that rotates about some point or axis, called a FULCRUM. The cave man is using a tree branch as a lever. The small stone is the _____.

fulcrum

2. The cave man exerts an EFFORT FORCE on the end of the lever. The large rock resists the effort force with a reaction force against the lever. This is called the RESISTANCE _____.

FORCE

3. In this lever system the effort force tends to move the lever in a *_____ direction. At the same time the resistance force tends to turn the lever in a *_____ direction.

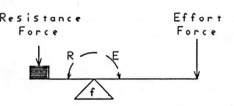

* clockwise or counter-clockwise

counter-clockwise

4. If the effort force rotates the lever about the fulcrum in a clock-wise direction, the turning effect of the effort force must be *_____ than the resistance turning effect.

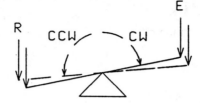

greater

   * greater or smaller

5. If the lever is balanced and is not turning, then we know that the turning effect of the effort force must be

*_____ turning effect of the resistance force.

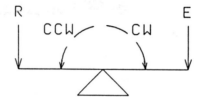

equal to

*less than, equal to, greater than

6. To study the turning effect of forces we need to know the line of action of the forces. We do this by extending the force arrows in the picture

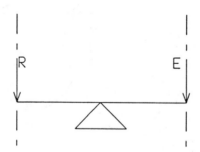

by using broken lines in both directions. The forces can be thought of as acting anywhere along these lines.

These are called the _____ of_____ .

lines of action

7. If a lever is balanced, we say that it is in EQUILIBRIUM. The turning effect of the effort force and the turning effect of the resistance force are equal and opposite in direction when the lever is in a state of _____.   equilibrium

8. This lever is in equilibrium. The forces E and R equal in magnitude. Their lines of action

*_____ equidistant from the fulcrum.   are

　　*are or are not

9. The turning effect of a force acting on a lever depends on two things:

   1. The magnitude of the _____.   force

   2. The _____ between the lever's fulcrum and the line of action of the force.   distance

10. This _____ effect of a force acting on a lever is called a MOMENT.   turning

11. When applied to bolts, nuts and screws the _____ is called TORQUE (pronounced TORK).   moment

12. Calling moments "clockwise" and "counter-clockwise" is rather cumbersome, therefore we assign them positive (+) and negative (-) signs. We do this so that we don't confuse the directions that the moments are trying to turn a body.

If we choose the + sign to represent all moments that are trying to turn the body in the clockwise direction, then all counter-clockwise moments will have a ____ ____. minus sign

Remember that the + and - signs don't mean that one moment is greater or less than the other. The signs only show the direction of _____. rotation

13. It is customary to use the positive (+) sign to show clockwise rotation. There is an exception to this rule when we are discussing pitching, yawing or rolling moments of aircraft. A nose-up pitching moment is always positive, a yaw to the right and a right wing down moment are also always positive _____. moments

14. To avoid confusion it is wise to show the direction of the positive moments in the upper right hand corner of your work paper by this symbol.

15. We said that to calculate the moment produced by a force that we must know the amount of the force and the distance from the fulcrum to the _____ of _____ of the force. line of action

16. The distance from the fulcrum must be the shortest distance to the line of action of the _____. force

17. This shortest distance is called the MOMENT ARM. It is found by drawing a line perpendicular (90° angle) to the line of action through the fulcrum. This line is called the _____ of the force.   moment arm

18. Which line in this picture shows the shortest distance from the line of action to the fulcrum? _____.   C

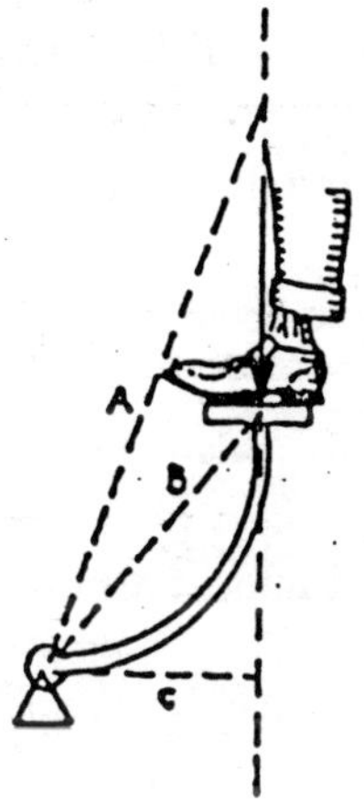

19. Which line makes a 90° angle with the line of action? _____.   C

20. Which line above represents the moment arm? _____.   C

21. In this lever system the moment at the fulcrum produced by force E is found by multiplying force E, by the length of the moment arm. The force E is _____.   2 lb.

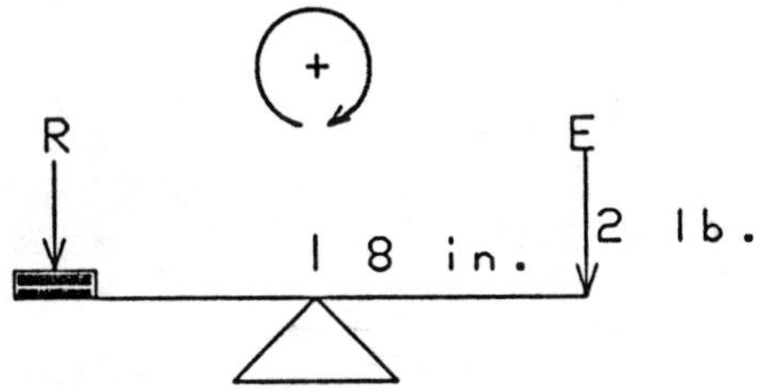

The length of the moment arm is _____.   8 in.

The moment is _____.   +16 inch-pounds.

Some scientists use the units of pound-inches so as not to confuse the units with energy or work units. Don't confuse moment units with work or energy

22. If the lever system shown above is in equilibrium the moment produced by force E (+ 16 in-lb) must be balanced by the moment produced by force ____.

R

23. What is the magnitude of the moment at the fulcrum produced by force R if the system is in equilibrium? _____.

-16 in-lb

24. FOR EQUILIBRIUM, THE SUM OF THE MOMENTS EQUAL ZERO.

This means that the positive moments and the negative moments must cancel each other. Since they cancel each other, their sum is ____.

zero

25. If the force R in question 21 is 4 lb., calculate what it's moment arm is for the system to be in equilibrium. Moment arm = ____ inches.

4

26. Let's look at three lever systems that are in equilibrium:

In lever system A shown here,

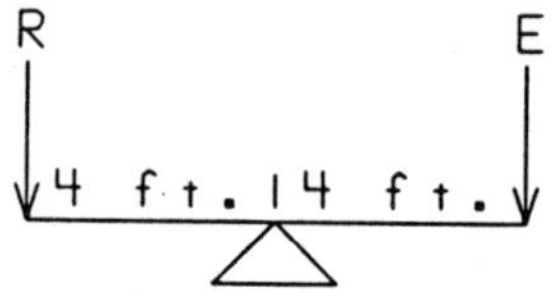

R _____ E. (how many times ?)

equals

In lever system B shown here,

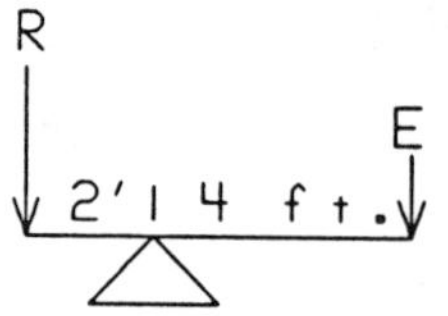

R _____ E. (how many times ?)

is twice

In lever system C shown here,

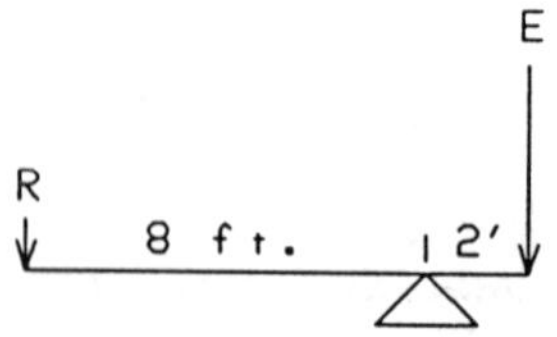

R _____ E. (how many times ?)

is one-quarter

27. In each of the above levers, the point where the fulcrum is located is called the CENTER OF GRAVITY (CG). The CG can be thought of as a point where all the weight of the body is concentrated. If the body is supported at its CG, the body will be *_____. in balance

*in balance or out of balance

28. For equilibrium to exist, the sum of the moments about the CG must be _____. zero

29. If a body is not in moment equilibrium, it will rotate about its _____. CG

30. Each of the above levers push down on the fulcrum with a force equal to ____ + ____. R + E

31. If the fulcrum is strong enough to resist the downward force, it will push back (react) with a force equal to ____ + ____. R + E

32. In addition to the sum of the moments being zero, another requirement must be met for equilibrium. For equilibrium, the sum of the FORCES must be ____. zero

33. For a body to be in a state of equilibrium:

THE SUM OF THE MOMENTS MUST BE ZERO AND THE SUM OF THE FORCES ACTING ON THE BODY IN ANY DIRECTION MUST BE ZERO.

All forces in any direction must _____. cancel out

34. Suppose we want to find the CG of this lever system.

We first put the fulcrum at some unknown distance, s, between the two forces. The force that the fulcrum reacts on the lever is ____ lb.

30

35. What we have just done is solve the equation:

Sum of the vertical forces = 0.

We use the Greek letter, Σ, (capital Sigma) to mean "sum of" and the symbol F_y to mean "forces in the vertical direction."

So: $\Sigma F_y = 0$.

There are no other forces in this problem so we have satisfied one of the requirements for _____.

equilibrium

36. Redrawing the lever system with the fulcrum reaction force shown as an upward force and showing our + moments, we have this picture.

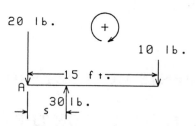

Clockwise moments are *_____.

+

     * + or -

38

37. In writing a moment equation it is not necessary to use the CG as the pivot point. You may assume that the system will rotate about any point. However, the distance to the CG must be measured from the assumed point of rotation.

Let's choose point A as the assumed pivot point. The distance to the CG must be measured from _____. point A

38. To write our "moment equation", we must find the moment that each force produces about our assumed point of rotation, A.

The 20 lb. force produces a moment about point A of:

Moment = force x moment arm = 20 lb. x 0 in. = _____ 0

Any time that the line of action of a force passes through the pivot point, its moment arm is ____ and so the 0

moment is ____. 0

39. Next we consider the 30 lb. force to determine what moment it produces about point A.

Moment = force x moment arm = 30 x s ft-lb.

The direction that the system would rotate about point A if it was acted upon by only the 30 lb. force is _____. counter-clockwise

According to our moment sign direction, this is a *_____ moment. −

* + or −

This moment is −30s.

40. The final moment is produced by the 10 lb. force It has a moment arm

(from point A) of _____ feet.   15

The moment = 10 x 15 = _____ ft-lb.   150

The direction that the system would rotate if it pivoted about point A and

was acted on by only the 10 lb. force is _____.   clockwise

So the moment is +150 ft-lb.

41. Next we must sum up all the moments and, as they must equal zero to

satisfy the equilibrium, set them equal to zero.:

$\Sigma M_A = 0$ $-30s + 150 = 0$

Solving this equation;   s = _____ ft. from A.   5

42. Find the Center of Gravity of this airplane.

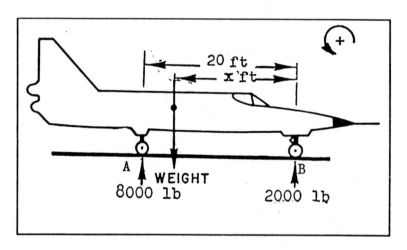

Step 1. What is the weight of the plane? _____.   10,000 lb.

Step 2. Write the number 10,000 lb. on the drawing for the weight.

Step 3. Choose a pivot point. Because the distance from the _____ wheel is shown on the drawing, it seems logical to choose the nose wheel as the _____.

nose

pivot point

We will write the moment equation about this point (name it point B).

Step 4. What is the moment of the 2000 lb. force about B? _____ ft-lb.   0

Step 5. What is the moment of the 10,000 lb. force about B? _____ ft-lb.   10,000x

Is this moment + or - ? _____.   +

Step 6. What is the moment of the 8000 lb. force about B? _____.   -160,000 ft-lb

Step 7. Write the entire moment equation:

$$\Sigma M_B = \underline{\hspace{2in}} = 0$$ 10,000x - 160,000

Step 8. Solve for x. x = _____.   16 ft.

43. Try the same weight and balance problem using point A as the pivot point. Write the entire moment equation:

$$\Sigma M_A = \underline{\hspace{2in}} = 0$$ -10,000(20-x) + 40,000

Solve for 20-x. 20-x = _____.   4 ft.

Is this the same CG location that we found above? _____.   yes

44. Does it make any difference where we choose the pivot point? _____   no

Must we measure all distances from the pivot point? _____.   yes

45. Solve this moment problem. Find the wheel reaction forces, F_m and F_n of the pictured 10,000 pound airplane.

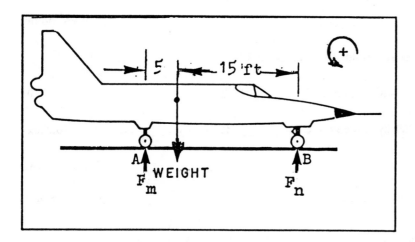

From the condition of equilibrium of forces, $F_m + F_n =$ _____ lb.   10,000

Write the moment equation about point A:

$\Sigma M_A =$ _____. $20F_n - 50{,}000 = 0$

Solve for $F_n =$ _____ lb.   2500

Solve for $F_m =$ _____ lb.   7500

46. If the engine of the plane in question 45 is running and is developing 2000 lb. thrust and the brakes are locked, find the new values of F_m and F_n. The thrust line of action is parallel to the runway and is 5 feet above the runway.

$F_m =$ _____ lb.   7000

$F_n =$ _____ lb.   3000

CHAPTER FIVE

NEWTON'S LAWS OF MOTION

1. Imagine a basketball sitting on a flat table. The ball will not move unless something is done to make it move. If we push, pull, hit or blow on the ball we can start the ball moving. We are creating an UNBALANCED FORCE on the ball.

Something that makes the ball begin to move is called an _____ _____.   unbalanced force

2. If we say that the ball is sitting still or if we say that the ball is moving at a certain velocity, we are describing its STATE OF MOTION. Whenever an object's state of motion is changed, we know that an

_____ _____ is acting on the object.   unbalanced force

3. An unbalanced force is defined as a push or pull that makes an object change its _____ of _____.   state of motion

4. All objects have a resistance to any _____ in their state of motion.   change

5. This unwillingness of an object to change its state of motion is called INERTIA. Because all objects have _____, they will not start to move by themselves.   inertia

43

6. Suppose that you are driving your car down the street and you come to a red traffic light. To change the state of motion of the car you must apply a braking _____ to the car. force

7. The inertia of the car not only resists starting but also resists stopping. Inertia resists any _____ of _____. change, motion

8. Suppose you are driving on an icy pavement and you try to turn a corner and skid sideways. Inertia also resists any change in direction of the car. It wants to *___ ___ _____ go straight ahead.

   * turn the corner. or, go straight ahead.

9. VELOCITY is a combination of speed and direction. If an object has either its speed or its direction of motion changed it will have its _____ changed. velocity

10. All objects have inertia and are unwilling to have their _____ changed. velocities

11. Because an object has inertia its velocity will not change unless some _____ force acts on it. unbalanced

12. Newton's first law of motion states that an object will not change its _____ unless some velocity force acts on it. unbalanced

13. If two equal and opposite forces act on a body at the same time, will the velocity of the body change? _____. No

This is why we say "unbalanced" force.

14. If you saw an airplane during takeoff and noticed that the speed of the plane was increasing, you could conclude that an _____ force was acting on it. unbalanced

15. You see another plane landing. The plane slows to taxi speed and turns off the runway onto a taxi strip. To slow the plane down after touchdown an _____ _____ was required. unbalanced force

16. To turn the plane from the runway at constant speed a change in direction and _____ was required. velocity

17. Newton's first law states that an object in motion will continue in motion without the continuous action of a _____ on it. force

18. If it was possible to remove all outside forces acting on a moving body, the body *_____ travel at an unchanging speed in a straight line forever. would

    *would or would not

19. Newton showed that an object's _____ would change only when an unbalanced force acted on it. velocity

20. Newton's first law also states that when an object is at rest that it will remain at rest and when a body is in motion it will remain in motion in a straight line unless it is acted upon by an _____ _____. unbalanced force

21. If an object's velocity is changing, it is ACCELERATING. We usually think of accelerating only as going _____ but this is not so. faster

45

22. In which of the below examples is the airplane accelerating? _____.   b

   a. A plane is flying in a straight line at an unchanging speed of 300 knots.

   b. A plane is flying in a circle at an unchanging speed of 300 knots.

23. An airplane during takeoff roll is accelerating.

   An airplane slowing down after landing is _____.   accelerating

   An airplane turning off the runway is _____.   accelerating

24. At the first of this chapter we talked about a basketball on a flat table

   and its unwillingness to move.

   Suppose that the basketball was filled with water instead of air. It would be

   *_____ to begin to move the ball.   harder

   *easier or harder

25. The water-filled ball would be more unwilling to have its state of motion

   changed. The _____ of the water-filled ball would be more.   inertia

26. Suppose it took ten times as much force to get the water-filled ball

   moving as it took to move the air-filled ball. We would say that the inertia

   of the water-filled ball was *_____ that of the air-filled ball.   ten times

           * how many times

27. The water-filled ball acted as if it contained a total of ten times as much matter as the air-filled ball. So we say that the MASS of the water-filled ball was _____ times the mass of the air-filled ball.

ten

28. We can say that mass is a measure of inertia. Because all objects have mass, they also have _____.

inertia

29. Mass and weight are not the same. The quantity of matter in an object is its *_____. The force of gravity acting on an object is its *_____.

mass, weight

    * weight or mass

30. Gravitational force acting on a mass determines its weight. An object with a certain mass may weigh 60 pounds on earth but only 10 pounds on the moon. The moon's gravity is *_____ that of the earth's.

1/6

    * six times, the same as, 1/6

31. The mass of the object on the moon will be *_____ its mass on earth.

the same as

    * six times, the same as, 1/6

32. Neglecting air resistance, all falling bodies near the surface of the earth will be accelerated by the gravitational force by the same amount. This acceleration is called the "gravitational constant", g.

g is an _____.

acceleration

33. The weight, W, of an object is equal to the gravitational constant multiplied by the mass, m, of the object.

$$W = (g)(m)$$

or $\quad m = \dfrac{W}{g}$

Mass is *_____ proportional to weight.   directly

 * directly or inversely

34. The acceleration of free falling bodies on earth is approximately 32 feet per second per second. This means that the body's vertical speed will increase by 32 feet per second for each second that it falls, if air resistance is neglected.

Accelerations are said to have the units of "feet per second squared" (fps^2).

This is easier to say than "feet per second per second."

The gravitational constant, g, has the value of _____ fps^2.   32

35. If an object on earth weighs 64 pounds its mass will be: $m = \dfrac{W}{g} = \dfrac{64}{32} =$ _____ units.   2

36. The units of mass are:

$$m = \dfrac{W}{g} = \dfrac{pounds}{feet\ per\ second^2} = \dfrac{pounds\text{-}seconds^2}{feet}$$

This is too complicated to say, so we give it a "nickname", SLUGS.

An object that weighs 96 pounds on earth has a mass of 3 _____.   slugs

48

37. The use of the term slugs for mass and pounds for weight may help in keeping the two concepts clearly separated. Mass is measured in _____ and weight is measured in _____.

slugs
pounds

38. To produce the <u>same</u> acceleration, a body with a larger mass requires *_____ force.

* more or less

more

39. The relationship between force and mass to produce the <u>same</u> acceleration can be stated as:

FORCE IS *_____ PROPORTIONAL TO MASS.

* DIRECTLY or INVERSELY

DIRECTLY

40. In symbol form this is written as: $F \sim m$

The symbol, $\sim$, means "proportional to". F stands for _____.

unbalanced force

41. The greater the mass of the object, the more inertia it has and the same force will produce

*_____ acceleration.

* more or less

less

42. We can give an object a certain acceleration by applying a _____ to it.

force

43. If we want to give the same object a greater acceleration, we must exert a _____ force to it.   greater

44. We can say, that if the mass of an object does not change, that:

   FORCE IS *_____ PROPORTIONAL TO ACCELERATION.   DIRECTLY

   * DIRECTLY OR INVERSELY

45. In symbol form this is written as: $F \sim a$

The symbol means _____ __.   proportional to

46. We can combine these two proportionality statements into a formula:

   $F = ma$ or $a = \dfrac{F}{m}$. This is called the force formula.

47. The force formula is Newton's second law. In words it is,

"IF A BODY IS ACTED UPON BY AN UNBALANCED FORCE, THE BODY WILL ACCELERATE IN THE DIRECTION OF THE FORCE AND THE ACCELERATION WILL BE DIRECTLY PROPORTIONAL TO THE UNBALANCED FORCE AND INVERSELY PROPORTIONAL TO THE _____ OF THE BODY."   MASS

48. There are several forces acting on this airplane.

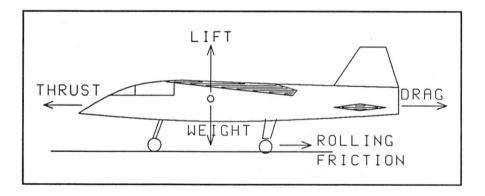

During takeoff, for instance, we have the thrust of a propeller (or jet thrust) acting to accelerate the airplane. We also have aerodynamic drag and rolling friction which we must overcome if the plane is to accelerate.

The NET ACCELERATING FORCE is the unbalanced force, F, which we use in the F = ma formula.

The net accelerating force is the thrust minus _____ and _____ _____.   drag
                                                                                                              rolling friction

48. Anytime we use the force formula we must remember that the symbol _____ means the resultant of all forces in the direction of motion.   F

49. An airplane weighs 24,000 pounds and has a turbojet engine which produces 12,000 pounds of thrust. The combined rolling friction and drag during takeoff are 2000 pounds and are considered to be constant during the takeoff run.

Find the following:

    a. The mass of the airplane, m = _____.   750 slugs

    b. The unbalanced force, F = _____.   10,000 lb.

    c. The acceleration, a = _____.   13.33 fps^2

50. Newton stated in his third law,

"FOR EVERY ACTION FORCE THERE IS AN EQUAL AND OPPOSITE REACTION FORCE."

Note that for this law to have any meaning, there must be an interaction between the force and the object. For example, in a turbojet aircraft engine the expanding gases exert a force to the rear of the plane. The equal and opposite reaction produces a thrust force in the _____ direction.   forward

51. Newton's three laws of motion are summarized as:

   1. The law of inertia.

   2. $F = ma$

   3. The law of action and reaction.

CHAPTER FIVE - MORE PRACTICE PROBLEMS

1. A pilot puts his 9600 pound airplane into a 35° dive as shown here.

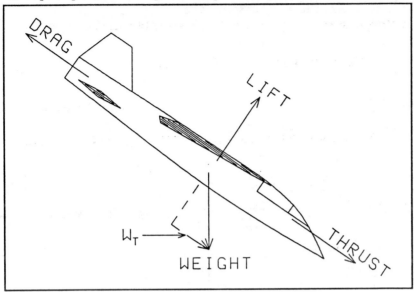

If the thrust and drag were equal at the point of entry into the dive, find the following:
    a. The component of weight in the direction of thrust.
       $W_T =$ _____ lb.   5506.3
    b. The unbalanced forward force. $F =$ _____ lb.   5506.3
    c. The plane's acceleration $a =$ _____ fps^2 18.35

2. A shuffleboard player accelerates his disk at the rate of 256 fps^2. The disk weighs 0.25 lb., the combined sliding friction and air drag is 0.165 lb., the cue stick makes a 30° angle with the ground. Find the force that the player pushes on the cue stick.   Force = _____ lb.   2.5

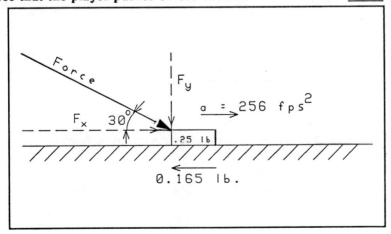

3. A pilot is flying straight and level at a constant airspeed. The thrust and drag are both 5,000 lb. The lift equals the weight of 10,000 lb. The pilot decides to climb and, at the same time, to accelerate to a higher speed. The climb is at an angle of 10° and the acceleration is 10 fps².

Find the thrust required at the start of this maneuver.

$T_R =$ _____ lb.

9861

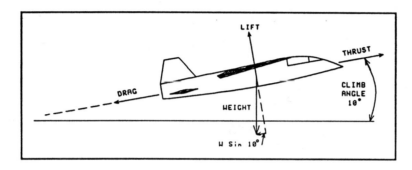

CHAPTER SIX

LINEAR MOTION

1. In Chapter Five we studied Newton's laws of motion but we didn't really find the relationships that exist among velocity, acceleration, time and distance. These will be investigated in this chapter.

We learned that in order for a change in the state of motion of an object to occur, an unbalanced _____ must act on the object. force

2. To observe the change in an object's state of motion, we must compare the object's state of motion after the change occurred, with its state of motion before the change occurred.

If we see an object moving steadily past us, we can't know that the object's state of motion has _____. changed

3. When we observe an object change from a state of zero motion to a moving state, we make two observations.

First, we observe it at rest, and then we observe it when it is moving.

We locate the object both in _____ and in space to see that its state of motion has changed. time

4. An airplane is observed on the runway with its brakes locked at 0800. At exactly 0801 you notice that the plane is moving down the runway. You are observing a _____ in the plane's state of motion. You have located the plane at <u>two points in time</u>, at 0800 and at 0801. change

55

5. At the first point or INSTANT in time, the plane was at rest (zero motion). The next time you looked, at the instant of time of _____, the plane was moving.

0801

6. So you know that, at some time between 0800 and 0801, a force must have been applied to the plane to make it _____ its state of motion.

change

7. Between the instant of time of 0800 and the instant of time of 0801, is a length or INTERVAL of time, of _____ minute.

one

8. Suppose that a race car is moving down a straight track at a uniform speed and we find that it travels one mile between 0930 and 0931. The car is moving uniformly at a rate of one mile in a time *_____ of one minute.   * instant or interval

interval

The car's speed is _____ miles per _____.

60, hour

9. To find how far the race car goes in any time interval, we simply use the following formula:

Distance, s, = *Speed, v,* × *Time Interval, t.*   or $s = vt$

10. The race car's speed is 60 mph. In 4 hours, how far does it travel?

$s = vt$ = _____ × _____ = _____ miles.

60 × 4 = 240

56

11. Motion in a straight line is called LINEAR motion. Motion along a curved path is called CURVILINEAR motion. We will only discuss linear motion in this chapter. If the race car is running on a straight track at constant speed, we say that it is in UNIFORM LINEAR MOTION.

So, if something is moving in uniform linear motion, it is moving in a straight line and its speed is _____.   constant

12. Suppose that the race car traveled in a straight line at a speed of 120 miles per hour for ½ minute and then stopped for ½ minute. The race car is not in ____ ____ ____.   uniform linear motion

13. Let's see how far the car traveled in the whole minute.

During the first ½ minute (1/120 hour) the speed was 120 mph.

   $s = vt = 120 (1/120) = 1$ mile

During the second 1/2 minute the speed was 0 mph.

   $s = vt = 0 (1/120) = 0$ miles

Total time interval was _____.   1 minute

Total distance traveled was _____.   1 mile

14. So, in spite of the fact that the car was stopped for ½ minute, it still traveled one mile in one minute time interval.

Its AVERAGE SPEED, v_{av}, was _____ mph.   60

57

15. Suppose that an airplane starts its takeoff roll at exactly 0800 and leaves the ground at exactly 0801. If the distance that the plane traveled before becoming airborne was 6000 feet, the average speed during the takeoff roll was: $s = (v_{av})(t)$ or $v_{av} = \frac{s}{t}$

Commonly we use the terms "velocity" and "average velocity". This is incorrect. The correct terms are "speed" and "average speed."

We must remember that speed is the magnitude of velocity but does not consider direction. In the above example: $v_{av} = \frac{6000 \text{ ft.}}{60 \text{ seconds}} = \underline{\qquad}$.   100 fps

NOTE: In these basic formulas we must use basic units.

Time is measured in seconds (sec). Distance is measured in feet (ft). Speed is measured in feet per second (fps). Acceleration is measured in feet per second per second (fps^2).

16. As we shorten the time interval to measure an instant (an infinitesimally small interval of time) and find the speed at this instant, we have found the INSTANTANEOUS SPEED of an object. The time interval approaches _____ seconds in determining instantaneous speed.   zero

17. The Greek letter capital DELTA, Δ, means "the change in". Δv means, the _____ in _____.   change in speed

18. The rate at which an object changes its speed is its _____.   acceleration

19. If we want to calculate the acceleration of an object, we need to know two things:

    1. the change in its speed.

    2. the length of the _____ interval in which the change took place.   time

20. If we find the object's acceleration over a time interval, we have found its *_____ acceleration.   average

    * instantaneous or average

21. To calculate an object's average acceleration, we must know two things:

    1. Δv (change in _____).   speed

    2. Δt (change in _____).   time

22. Average acceleration, a_{av}, is the change in speed per change in time.

$$a_{av} = *\underline{}$$ $\dfrac{\Delta v}{\Delta t}$

    * write the symbols

23. Suppose that a car driver notices his speed is 30 mph (44 fps) and he starts a stop watch at this instant. Ten seconds later he reads 60 mph on his speedometer.   His Δv was _____ fps.   44

                                    His Δt was _____ sec.   10

24. Using the acceleration formula, his average acceleration, a_{av} = _____ fps^2 4.4

25. If the car had a constant acceleration during the run, then the instantaneous acceleration is *_____ the average acceleration.

* different than or the same as

the same as

26. If the accelerating force on an object is constant, the object will be UNIFORMLY ACCELERATED. Uniform acceleration occurs when a _____ unbalanced force is applied to an object.

constant

27. The formulas of motion are simplified if we restrict our discussion to uniform acceleration. The following discussion is confined to considering the acceleration as being a _____.

constant

28. If the acceleration is now a constant, we can eliminate the symbol a_{av} and simply use ____.

a

29. The acceleration formula now is:

$$a = \frac{\Delta v}{t} = \frac{v - v_o}{t}$$

Where: v_o = original speed of the object

v = final speed of the object

t = time *_____ that acceleration acted

* interval or instant

interval

30. Rearranging the terms leads to: $v = at + v_o$

31. If a car is traveling at 30 mph (44 fps) and the driver accelerates the car at a uniform rate of 8 fps² for 5.5 seconds, find the final speed.

v = _____ fps, _____ mph.   88, 60

32. A pilot is flying his airplane at a speed of 100 k. He opens the throttle and accelerates the plane and after a time interval of 30 seconds the airspeed is 150 knots. Find the acceleration of the airplane. (knots x 1.69 = fps)

a = _____ fps².   2.82

33. Later, the pilot in the above problem reduces the plane's airspeed. If the deceleration (negative acceleration) is 4 fps² and the plane slows down for 20 seconds, find the final airspeed. v = _____ fps, _____ mph.   173.5, 118.3

34. An airplane on its takeoff run passes an observer at a speed of 50 knots. Assume a constant acceleration of 4 fps² and a takeoff speed of 95 knots. Find the time interval from the time the airplane passed the observer until takeoff. t = _____ seconds   19.01

35. We found (in question 15) that the distance that an object moved depended on the average speed and the time interval involved.

$$s = (v_{av})(t)$$

The average speed is: $$v_{av} = \frac{(v + v_o)(t)}{2}$$

Substituting values: $$s = \frac{(v + v_o)(t)}{2}$$

We also found (in question 30) that $v = at + v_o$.

Substituting this value of v, we find:

$$s = v_o t + \frac{1}{2}at^2$$

This formula can be used to find a, v_o, or a, but should not be used to find _____.

t

36. A race car passes the starting line at 100 fps and accelerates at a uniform rate of 5 fps². The car crosses the finish line 4 seconds later.

Find the length of the track. s = _____ feet.

440

37. An airplane passes an observer at 20 knots during its takeoff run. The plane is seen to leave the ground 1000 feet beyond the observer. The time interval between the two observations was exactly 10 seconds.

Find the constant acceleration. a = _____ fps²

13.24

38. Distance, uniform acceleration, and speed can be directly related in a formula by substituting the value of:

$$t = \frac{v - v_o}{a}$$

into the formula derived in question 35. The new formula is:

$$s = \frac{v^2 - v_o^2}{2a}$$

Note that there is no time, t, symbol in this equation.

It *_____ necessary to know time to solve for takeoff distance.   is not

  * is or is not

39. An airplane touches down upon landing and decelerates at a constant rate of 5 fps². The landing roll-out takes 1250 feet.

Find the touchdown speed. v_o = _____ fps, _____ knots   111.9, 66.2

40. Cargo is dropped from a hovering helicopter and reaches the ground with a vertical velocity of 80 fps. Neglecting air resistance, find the height of the helicopter.

Hint: Free falling bodies are accelerated by gravity.

$g = 32$ fps² (on earth).

s = _____ feet.   100

41. A helicopter moves forward from a hover to a speed of 60 knots while covering 1000 feet ground distance.

Find the uniform acceleration. a = _____ fps²   5.1

42. Find the speed of the above helicopter after it has traveled 500 feet ground distance. v = _____ fps, _____ knots.   71.7, 42.4

43. A race car with a uniform acceleration of 10 fps² passes the starting line at a certain speed and accelerates for 1000 feet. At the end of the acceleration period it has attained a speed of 150 mph.

Find the speed of the car as it passed the starting line. (mph x 1.47 = fps)

v = _____ fps, _____ mph.   169.2, 115.1

SUMMARY OF CHAPTER SIX FORMULAS

$v = at + v_o$ v_o = original speed, fps

v = final speed, fps

$s = v_o t + \frac{1}{2}at^2$ a = acceleration, fps²

t = time interval, seconds

$s = \dfrac{v^2 - v_o^2}{2a}$ s = distance, feet

CHAPTER SIX - MORE PRACTICE PROBLEMS

1. An airplane weighs 32,000 pounds and has a turbojet engine which produces 12.000 pounds of thrust during takeoff. The combined rolling friction and drag are 2,000 pounds during takeoff. All forces are assumed to be constant during the takeoff run.

Find:

Acceleration during takeoff. a = _____ fps^2.   10

If the airplane starts from brakes locked and takes off at 150 knots, Find the takeoff distance.

s = _____ feet.   3213

Find the time interval for the takeoff.

t = _____ seconds.   25.35

2. Suppose that you jump off a 3 foot high table and land stiff-legged on a concrete floor. If your stopping distance is 0.03 feet, how many "G" units does your body experience?

Hint: 1 "G" unit is equivalent to an acceleration of 32 fps^2.

Gs = _____ .   100

3. A bomb is dropped from a bomber flying horizontally and strikes the ground in 10 seconds.

Neglecting air resistance, find the vertical velocity of the bomb at impact.

v = _____ fps.

320

4. How high was the bomber in problem 3?

s = _____ feet.

Check your answer by using a second equation and comparing your answers.

Both of

your

answers

should

agree.

CHAPTER SEVEN

CURVILINEAR MOTION

1. The velocity of a body moving in a curving path (curvilinear motion) at constant speed *_____ changing.   is

   * is or is not

2. Either a change in speed or a change in direction (or both) will change the

   _____ of a moving body.   velocity

3. An object moving in a curvilinear path must have a _____ acting on it.   force

4. If you swing a ball on a string around your head, as shown in the sketch, you *_____ exert a force on the ball to make it follow the circular path.   must

   * may or must

5. In the above sketch vector _____ represents the force you must exert to keep the ball circling.   A

6. This force is called CENTRIPETAL force. Centripetal means "acting toward the center." The ball tries to follow a straight line path, but the _____ force acting through the string pulls the ball toward the center and keeps it in a circular path.   centripetal

67

7. Newton's second law tells us that an unbalanced force on an object produces an acceleration on an object and that the acceleration will be in the direction of the _____.   force

8. The centripetal force on the ball in circular motion is toward the _____ of the circle.   center

9. The ball must then be _____ toward the center of the circle.   accelerating

This is called RADIAL ACCELERATION and the symbol a_r is used for it.

10. The velocity of an object moving in a curvilinear path is called its TANGENTIAL VELOCITY. The symbol v_t is used for _____ _____.   tangential velocity

11. The ball in circular motion will have a constant v_t if you swing it at *_____ revolutions per minute.   constant

* constant or varying

12. The tangential velocity, v_t, of an object being rotated depends upon:

    1. the radius of rotation, and

    2. the speed at which it rotates.

The formula for tangential velocity is:

$$v_t = \frac{\text{Revolutions Per Minute (RPM)} \times \text{Radius, ft. (R)}}{9.55}$$

$$v_t = \frac{(RPM)(R)}{9.55}$$

This formula is useful in computing propeller or helicopter rotor tip speeds.

13. If a helicopter has a 19.1 ft. blade radius turning at 300 RPM, find the tip speed. $v_t = $ _____ fps. 600

14. The radial acceleration of an object in rotation depends on:

    1. its tangential velocity, and

    2. its radius of rotation.

$$a_r = \frac{v_t^2}{R}$$

15. For the above helicopter, find the radial acceleration.

$a_r = $ _____ fps^2 18,848

16. The centripetal force on the string can also be found by substituting the value of acceleration into Newton's force formula, $F = ma$.

$$CF(unbalanced\ force) = ma_r$$

substituting: $m = \dfrac{W}{32}$ into the above and $v_t = \dfrac{(RPM)(R)}{9.55}$ into $a_r = \dfrac{v_t^2}{R}$

and then into the CF equation above, we get:

$$CF = \frac{(W)(R)(RPM^2)}{2918.48}$$

17. To find the centripetal force on the string of the orbiting ball, we must know the weight of the ball, the RPM and the _____ of the string. length

18. If the string is 3 feet long, the ball weighs 1 pound and you spin it around 100 times in one minute, find the tension on the string. CF = _____ lb. 10.28

19. If the string breaks, the centrifugal force will no longer pull the ball in a circular path. The ball will fly off in a straight line in the *_____ direction. tangential

* radial or tangential

20. From Newton's third law (action and reaction) we know that the ball must exert an equal and _____ reaction on the string. opposite

21. CENTRIFUGAL means "flight from the center." The centrifugal force is directed away from the center of rotation.

Vector ____ represents the centrifugal force. B

22. If a car is going around a flat curve, friction between the road and the tires exerts a _____ force on the car and makes it turn from its linear path. centripetal

23. The driver will be forced to the outside of the turn by _____ force. centrifugal

24. If the friction between the road and the tires is not enough, the car will continue its _____ motion and it will skid. linear

25. There is much confusion between centrifugal and centripetal force.

_____ force is required to change the linear path into a curvilinear path.　　　　Centripetal

_____ force is only the reaction force.　　　　Centrifugal

SUMMARY OF CHAPTER SEVEN FORMULAS

$v_t = \dfrac{(RPM)(R)}{9.55}$　　　　v_t = tangential velocity, fps

$CF = \dfrac{W v_t^2}{32R}$　　　　RPM = revolutions per minute

$CF = \dfrac{WR(RPM^2)}{2918.48}$　　　　R = radius of rotation, ft.

$a_r = \dfrac{v_t^2}{R}$　　　　CF = centripetal force, lb.

　　　　　　　　W = weight, lb.

　　　　　　　　a_r = radial acceleration, fps^2

CHAPTER SEVEN - MORE PRACTICE PROBLEMS

1. A turbine in a jet engine has a diameter of 22.9 inches. Find the tip speed of the turbine blades if the RPM is 10,000.

$v_t =$ _____ fps   999.13

2. If the above turbine throws a blade which weighs 0.1 lb., how much unbalanced force is created?

$CF =$ _____ lb.   3269.40

3. The OH-6A helicopter has a rotor diameter of 26.33 ft. and operates at 470 RPM. If the tip speed is restricted to M = .75 and the speed of sound is 1100 fps, find the limiting speed of the helicopter.

NOTE: Tip Mach Number, $M = \dfrac{v_t}{a}$. Where a = speed of sound.

$v_{max} =$ _____ fps = _____ knots.   177.09, 104.79

CHAPTER EIGHT

ENERGY

1. ENERGY is required to do work. If you have a lot of _____ you can do a lot of work.   energy

2. There are many types of _____. To name a few, there is atomic energy, chemical energy, solar energy, thermal energy and mechanical energy.   energy

3. In considering aircraft in flight, we are primarily interested in _____ energy.   mechanical

4. To wind a Cuckoo clock, you raise the weight that runs the clock. You have transferred some mechanical _____ to the weight.   energy

5. For the rest of the day the STORED energy was slowly released and did _____ as it ran the clock.   work

6. Stored energy in physics is given a special name. _____ energy is called POTENTIAL ENERGY.   stored

7. Stored energy is called _____ energy because it has the ability or "potential" of doing _____.   potential
    work

8. In mechanics the most common type of _____ energy is similar to the raised weight of the clock.   potential

It is properly called POTENTIAL ENERGY OF POSITION.

In this discussion we will shorten this and simply call it "potential energy" (PE).

9. If you raise a 10 lb. weight to a height of 5 ft., how much energy did you transfer to the weight? _____ ft-lb. 50

How much PE did you add to the weight? _____ ft-lb. 50

10. To measure the PE of an object, you need to know:

    a. the _____ of the object. weight

    b. the _____ of the object. height

11. The formula is simply: *PE = Wh*.

W is the _____ of the object and h is the _____ to which it is raised. weight, height

12. Suppose a pile driving machine has a 1000 lb. hammer and the hammer is raised to a height of 15 feet above the pile. What is its PE? _____ ft-lb. 15,000

13. If the pile driver hammer is released and strikes the pile and drives it into the ground. At the instant of contact, what is the PE of the hammer? _____. 0 ft-lb

14. Even though the pile driver hammer didn't have any PE when contact was made, it did have some mechanical energy of another kind. The energy that the pile driver hammer used to do its work was its _____ of motion. energy

15. This form of mechanical energy is called KINETIC ENERGY (KE).

Potential energy is stored energy of _____.   position

Kinetic energy is the energy of _____.   motion

16. A bowling ball scatters the pins because it has _____ energy.   kinetic

17. While the pile driver hammer was falling, did it have kinetic energy? ___   yes

18. As long as the pile driver hammer was at some height above the pile, did it have any potential energy? ___   yes

19. A falling body has both _____ energy of position and _____ energy of motion.   potential, kinetic

20. While the pile driver hammer was falling, faster and faster, its PE was changing to ___.   KE

21. During the time the hammer was falling its KE was *_____   increasing

while its PE was *_____.   decreasing

    * decreasing or increasing

22. The KE of an object depends on its speed and its mass. A bowling ball rolling at 50 fps will scatter the pins *_____ than if it is rolling at 25 fps.   more

    * more or less

23. A 16 pound bowling ball will scatter the pins _____ than a 12 pound ball rolling at the same speed.   more

24. The formula for calculating kinetic energy is: $KE = \frac{1}{2}mv^2$ where v is the speed (fps) of the object and m is the _____ of the object.   mass

25. The 16 lb. bowling ball has a mass = 16/g = _____.   0.5 slugs

What is its KE if it is rolling at 50 fps. KE = _____.   625 ft-lb

26. Suppose that a hovering helicopter is carrying a 320 pound load on a sling. The pilot releases the load and it falls to the ground in 2 seconds.

Before the load was released it had _____ energy but no _____ energy.   potential, kinetic

27. Find the speed of the above load as it hits the ground (ignoring air resistance) using the formula: $v = at + v_o$. v = _____.   64 fps

28. The load has a mass of _____.   10 slugs

29. The KE at impact is _____ ft-lb.   20,480

30. How much PE did the load have at impact? _____.   0 ft-lb

31. Before the load was released how much KE did it have? _____.   0 ft-lb

32. During the fall, all of the PE of the load before release was transformed into _____.   KE

33. The LAW OF CONSERVATION OF ENERGY says that energy can not be created nor destroyed but can change in form.

The potential energy of the load before release is *_____ the kinetic energy of the load at impact.   equal to

* less than, equal to, greater than

34. The numerical value of the KE at impact was found to be 20,480 ft-lb. What must the PE of the load have been before release? _____ ft-lb.   20,480

76

35. From the formula: *PE = Wh*, we can find the height of the above helicopter to be _____. 64 ft.

36. At any instant during the drop the TOTAL ENERGY (TE) of the load was _____ ft-lb. 20,480

37. A car is moving along a flat roadway at 100 fps speed. The car weighs 3200 pounds. Find the following:

    a. the mass of the car = _____ slugs. 100

    b. the KE of the car = _____ ft-lb. 500,000

    c. The PE of the car = _____ ft-lb. 0

    d. The TE of the car = _____ ft-lb. 500,000

38. Assume that there are no friction forces nor aerodynamic drag on the above car. With the engine turned off, what will the speed of the car be? _____. 100 fps

39. Under the above conditions, the car goes up a hill until it runs out of speed and stops. The driver puts the brakes on to keep the car from rolling backward. Find:

    a. velocity = _____. 0 fps

    b. KE = _____. 0 ft-lb

    c. TE = _____ ft-lb. 500,000

    d. PE = _____. 3200h

    e. height above roadway = _____. 156.25 ft.

40. Suppose that the car in Problem 37 and 38 was going up the hill and was just passing the 100 foot elevation point on the hill. Find the following:

    a. velocity = _____.   60 fps

    b. KE = _____ ft-lb.   180.000

    c. TE = _____ ft.-lb.   500,000

    d. PE = _____ ft-lb.   320,000

41. A mobile aircraft arresting gear has an energy absorbing capacity of 10,000,000 ft-lb. What is the maximum speed that a 16,000 pound aircraft can be arrested? v = _____ fps or _____ knots. (k x 1.69 = fps)   200, 118.2

SUMMARY OF CHAPTER EIGHT FORMULAS

$PE = Wh$ PE = Potential Energy, foot-pounds

$KE = \frac{1}{2}mv^2$ W = Weight, pounds

$TE = PE + KE = Constant$ h = height, feet

             KE = Kinetic Energy, foot-pounds

             m = mass, slugs

             v = speed, feet per second

             TE = Total Energy, foot-pounds

CHAPTER NINE

WORK and POWER

WORK

1. In science, only the jobs that require the use of energy are called _____. *work*

2. Pushing a car, lifting a weight, and holding a wrench are all jobs, but not all of them require doing _____. *work*

3. The landing gear supports the weight of an airplane. It does not use any energy in doing it, so it does no _____ on the airplane. *work*

4. The landing gear does exert a force on the airplane, but just exerting a _____ is not necessarily the same thing as doing work. *force*

5. Forces do not always use energy. A spring that is compressed and squeezed into a box will exert a force against the ends of the box. But the spring uses no _____. *energy*

and therefore does no _____. *work*

6. If you pushed all day against an unmoving wall, you would not be doing any work on it. You would get tired, because energy would be used up inside your body, but none of this energy would be used up on the wall. So no _____ would be done on the wall. *work*

7. If you know that a force is acting on an object, can you always know that work is being done on it? _____ .   no

8. All objects have INERTIA, or resistance to having their velocity changed. To overcome the _____ of a landing airplane, a _____ must be exerted to to slow the plane.   inertia, force

9. Whenever the velocity of an object is changing, you know that a force is acting on the object. During the time that the velocity is changing, energy is being transferred to the object. So, _____ is being done on the object.   work

10. Suppose you ran out of gas and tried to push your car, but didn't succeed. Your force wasn't enough to move the car any distance. Although energy was being used up inside your body, no _____ was transferred to the car because you did not succeed in moving it any _____ .   energy
distance

11. If a force is to do any work on an object, the force must move the object through some _____ .   distance

12. When a force moves a load through a _____ , energy is used in moving the load and _____ is done.   distance
work

13. Three things happen when work is done on an object:

    a. a _____ is exerted on the object.   force

    b. _____ is transferred to the object.   energy

    c. the object is moved a certain _____.   distance

14. The amount of energy transferred while doing work depends upon the amount of _____ exerted and the _____ that the object is moved.   force, distance

15. In the U.S. system we measure the distance in _____ and the force in _____.   feet / pounds

16. The unit of _____ is therefore called the **FOOT-POUND** (ft-lb).   work

17. The formula to find the work done when a force moves an object through a distance is:

WORK = FORCE (F) x DISTANCE (s)

The amount of work required to raise a 4000 pound helicopter to a 10 foot hover is _____ ft-lb.   40,000

18. If a boy pushes sideways on a sled with a ten pound force, but the sled doesn't move sideways in the direction of his push, does he do any work? ___   no

19. Suppose that the sled is sliding forward down a hill while the boy is pushing sideways. The sideways distance that the sled moves is zero, so the work that the boy does on the sled is _____.   0

20. For work to be done, a force must move an object in the direction of the _____.   force

21. Our work formula must be changed to include this concept of direction:

WORK = F x s(in the direction of the force)

22. Another way of saying this is, "only the amount of force in the direction of movement does any _____."   work

23. Suppose a 100 lb. force acts as shown below and moves a box 10 feet horizontally.

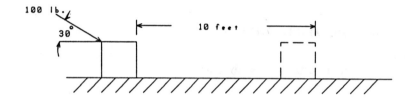

Is the work done on the box 1000 ft-lb? _____.   no

24. Only the horizontal component of the force acts in the direction of movement. Only the horizontal component does any work.

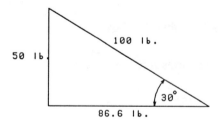

So the work done is _____.   866 ft-lb

82

POWER

25. Besides knowing the amount of work that is done, we often need to know the time that it takes to do a certain amount of work. The amount of work done in a certain interval of time is called the RATE of doing _____. work

26. Work is measured in units of force and distance (foot-pounds). Rate of doing work also involves the _____ interval. time

27. The rate of doing work is called POWER. The basic units of power are "foot-pounds per second." If it takes 2 seconds to do 4000 ft-lb of work, the POWER REQUIRED, P_R, is _____ ft-lb per second. 2000

28. If the big motor can do 10 ft-lbs of work per second and the small motor can do 5 ft-lbs per second, the _____ produced by the big motor is two times that of the small motor. power

29. James Watt was a Scottish mining engineer. The coal from the mines was raised up the mine shaft by horses. Watt perfected the steam engine for use to replace the horses and introduced the term HORSEPOWER (HP) based upon the amount of work a horse could do in a certain interval of _____. time

30. One horsepower is defined as being equal to 550 foot-pounds of work per second (33,000 ft-lb/min.) Thus if an engine does 5500 ft-lb of work in one second, we know the power of the engine in horsepower units is _____. 10

31. Work = _____ x _____ (in the direction of force).   force x distance

32. Power = Work divided by _____.   time

and Work = force x distance so, Power = <u>force x distance</u>
                                              time

but, distance = _____   speed
                  time

so, Power = _____ x speed.   force

33. To convert this equation to one involving horsepower units, we divide both sides by 550 (ft-lb/sec.).   HP = Power/550 = Force x Speed/550

In this equation, force is measured in _____.   pounds

Speed is measured in _____.   fps

Using the symbol v for speed, HP for Horse Power, and F for force:

$$HP = \frac{Fv}{550}$$

34. A jet airplane requires 10,000 pounds of thrust (force) to propel the plane at a speed of 1100 fps. What is the equivalent horsepower of this jet? _____.   20,000

35. If the speed of the airplane is measured in knots (nautical miles per hour), it is more convenient to use knots rather than fps in the equation. The conversion factor is: knots x 1.69 = fps.

The HP formula then becomes: $HP = \frac{Fv_{fps}}{550} = \frac{Fv_k}{325}$

What is the horsepower equivalent of an 8000 pound thrust jet engine, if the airplane's speed is 650 knots? _____.   16,000

36. There is much confusion between horsepower units and thrust units. The conversion factor formula must be used in relating the two. Horsepower is used for propeller aircraft and helicopters while thrust units are used for jet aircraft and rocket engines.

The formula can be rewritten to convert horsepower units to equivalent thrust as: $T = F = \dfrac{325 \; HP}{v_k}$

An aircraft has a reciprocating engine and propeller combination that produces 500 horsepower. How much thrust will the aircraft develop if it is flying at 162.5 knots? T = _____ lb.   1000

37. Calculate the horsepower available (P_A) from a 15,000 lb. thrust turbojet engine at:

    a. 160 knots. _____ HP.   7384.6

    b. 400 knots. _____ HP.   18,461.6

    c. 0 knots. _____ HP.   0

38. Calculate the thrust available, (T_A), from a 2500 HP reciprocating engine at:

    a. 150 knots. _____ lb.   5416.7

    b. 300 knots. _____ lb.   2708.3

    c. 20 knots. _____ lb.   40,625

SUMMARY OF CHAPTER NINE FORMULAS

$WORK = Fs$ $\qquad F$ = Force, pounds

$POWER = \dfrac{WORK}{t}$ $\qquad s$ = distance, feet

$POWER = \dfrac{F \times s}{v}$ $\qquad t$ = time, seconds

$POWER = F \times v$ $\qquad v_{fps}$ = speed, feet per second

$HP = \dfrac{Fv_{fps}}{550}$ $\qquad v_k$ = speed, knots

$HP = \dfrac{Fv_k}{325}$ $\qquad HP$ = Horse Power units

CHAPTER TEN

FRICTION

1. FRICTION is a resistance to movement between surfaces that are in contact with each other. Friction forces cause some of a machine's input energy to be wasted. The dragging force of _____ is at work.   friction

2. If we examined the surface of a material through a magnifying glass, we would see that even the smoothest of surfaces is really very rough. These rough surfaces lock together when they touch and _____ forces result.   friction

3. According to the LAW OF CONSERVATION OF ENERGY, friction energy is transferred to other kinds of energy. Friction forces cause

   *_____ energy.   wasted

   * wasted or destroyed

4. If you rub your hands together rapidly, you will feel them getting hot. The wasted energy has been transferred to _____ energy.   heat

5. If you raise a weight by using a pulley and you hear the pulley squeak, you know that some of the input energy is being transferred into _____ energy.

sound

6. In addition to the friction that comes from two surfaces sliding past each other (sliding friction), we also have friction between a round body rolling on a solid surface. This kind of friction is called _____ friction.

rolling

7. Not all friction is considered to be undesirable. We try to reduce the effects of friction in machines, but suppose you were driving on an icy road. To start or stop the car you *_____ want high friction forces.

would

    * would or would not

8. In a drag race the drivers spin their wheels before the race. The special rubber in their tires gets sticky when hot. Friction is *_____.

increased

    * increased or decreased

9. Another way that drag racers could increase the friction forces on the driving wheels would be to add weights over the driving wheels. More weight *_____ friction forces.

increases

    * decreases or increases

10. For the same surfaces in contact with each other, it has been found that the friction forces are proportional to the forces squeezing the surfaces together. This is called the NORMAL (perpendicular) force between the surfaces.

Another way of saying this is: $\dfrac{Friction\ Force}{Normal\ Force} = Constant$

If you double the normal force, the friction force is *_____. double

* one half, double, the same

11. The constant that relates the friction forces to the normal forces is called the COEFFICIENT of FRICTION.

Friction Forces and Normal Forces are measured in pounds (in the USA).

The dimensions of the coefficient of friction are _____. dimensionless

We use a lot of coefficients in engineering. All are dimensionless.

12. The lower case Greek letter, μ, (mu) is used to designate the coefficient of friction. F_f is used as the symbol for friction force and N is used for the symbol for normal force. Write the symbol equation for the relationship between these factors. μ = _____. $\dfrac{F_f}{N}$

13. The value of the coefficient of friction of aircraft tires on a runway depends upon such factors as: runway surface material and condition, rubber composition, tread design and wear, inflation pressure, and amount of braking pressure.

The coefficient of friction would be *_____ on a wet runway than on a dry one.   * more or less   less

14. When an airplane's tires are rolling on a runway without use of brakes, the friction force will be simply the rolling resistance. The rolling friction is a force that acts *_____ on the aircraft.   backward

   * forward or backward

15. By applying brakes the coefficient of friction is *_____.   increased

   * increased or decreased

16. If the brakes are applied hard enough they will lock and the braked wheels will stop rotating. When this happens the tires skid and the rubber is heated and reverts to its natural state and the coefficient of friction is

_____.   decreased

Braking distance is *_____ when aircraft brakes are locked (100% slip).   increased

   * increased or decreased

Many modern aircraft are equipped with non-skid brakes so that it is impossible to skid the tires.

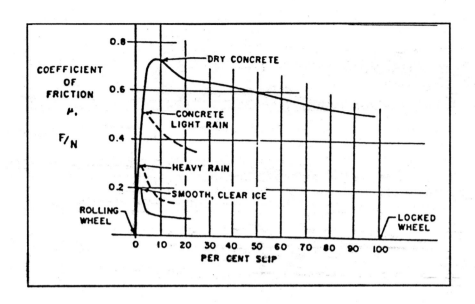

17. The above figure shows typical values of coefficient of friction to percent of brake slippage. Assume that a 10,000 lb. airplane is on a dry concrete runway and that 90% of the weight is on the main wheels. The normal force on the braked wheels is then 9,000 lb.

From the figure the maximum coefficient of friction is found to be about 0.7. The friction force is called the braking force F_b.

Find the maximum braking force on the airplane. $F_b =$ _____ .   6300 lb.

18. Under the same conditions, what would the braking force be if the pilot locked his brakes? $F_b =$ _____ .   4500 lb.

19. During a heavy rain storm, what is the maximum braking force that could be obtained? _____ .   2700 lb.

SUMMARY OF CHAPTER TEN (and earlier) FORMULAS

$$F_b = \mu N$$ F_b = Braking force, pounds

$$F = ma$$ μ = coefficient of friction

$$s = \frac{v^2 - v_o^2}{2a}$$ N = Normal (squeezing) force

$$CF = \frac{W(v_t^2)}{32R}$$

CHAPTER TEN - MORE PRACTICE PROBLEMS

1. The airplane in problem 17 above was moving at 100 knots when the brakes were applied and the average aerodynamic drag was 700 pounds. (Knots x 1.69 = fps.)

Find the total retarding force. F = _____. 7000 lb.

Find the deceleration. a = _____. -22.4 fps^2

Find the stopping distance. s = _____. 637.5 ft.

2. A 10,000 pound airplane is turning off the runway onto a taxi strip. The radius of turn is 100 ft. The plane is traveling at 25 knots. The coefficient of friction is 0.5. Calculate the forces to determine whether the plane will skid or not.

Hint: Find the centripetal force and compare it to the friction force.

CF = ____.   5578.3 lb.

F_f = ____.   5000 lb.

SKID? ____.   Yes

3. A car has 4 wheel brakes and is traveling at 100 fps when the driver sees an accident ahead and makes a panic stop. The coefficient of friction with locked brakes is 0.5.

Find the following:

   a. the normal force, N, on the braking wheels (in terms terms of the weight of the car).

N = _____ .   W

   b. the braking force, F_b = _____ .   .5W

   c. the mass of the car (in terms of W). m = _____ .   W/32

   d. the deceleration of the car. a = _____ .   -16 fps^2

   e. the stopping distance. s = _____ ft.   312.5